George A Shove

Life Under Glass

Containing suggestions toward the formation of artificial climates

George A Shove

Life Under Glass
Containing suggestions toward the formation of artificial climates

ISBN/EAN: 9783337418977

Printed in Europe, USA, Canada, Australia, Japan

Cover: Foto ©berggeist007 / pixelio.de

More available books at **www.hansebooks.com**

LIFE UNDER GLASS.

CONTAINING SUGGESTIONS TOWARD

THE FORMATION OF ARTIFICIAL CLIMATES.

BY

GEORGE A. SHOVE.

" If all were free,
Who would not, like the swallow, flit, and find
What season suited him? — in summer heats
Wing northward, and in winter build his home
In sheltered valleys nearer to the sun."

BOSTON:

JAMES R. OSGOOD AND COMPANY,

(LATE TICKNOR & FIELDS, AND FIELDS, OSGOOD, & CO.)

1874.

BOSTON:
STEREOTYPED AND PRINTED BY RAND, AVERY, & CO.,
117 FRANKLIN STREET.

DEDICATION.

———◆———

ONE of the foremost of English medical writers, Dr. James Johnson, emphatically says, "I declare my conscientious opinion, founded on long observation and reflection, that if there was not a single physician, surgeon, apothecary, chemist, druggist, or drug, on the face of the earth, there would be less sickness and less mortality than now obtains." And Prof. Magendie is reported to have addressed his students at the Medical College in Paris to the following effect: "Gentlemen, medicine is a great humbug. I know it is called a science. Science indeed! — it is nothing like science. Doctors are mere empirics when they are not charlatans. We are as ignorant as men can be. Who knows any thing in the world about medicine? Gentlemen, you have done me the honor to attend my lectures; and I must tell you frankly, that I know nothing about medicine. True, we are gathering facts every day. We can produce typhus-fever, for example, by injecting a certain substance into the veins of a dog; we can alleviate diabetes; and I see distinctly, *we are fast approaching the day when phthisis can be cured as easily as any disease.* But I repeat it to you, there is no such thing now as medical science. I grant you, people are cured; but how? Nature does a great deal; imagination does a great deal; doctors do — devilish little."

PREFACE.

To the man or woman who is blessed with
even the smallest of conservatories or green-
houses, and whose home is on the shady
side of the fortieth parallel of latitude, the
motive of the following little work will need
no apology. It is such a supreme satisfac-
tion to have a few hundred cubic feet of
space fenced in with crystal from the raging,
stinging winds of a Northern winter, — a sheltered nook where one can easily fancy it the
middle of May, while out of doors the frozen
blood of St. Januarius has not yet begun to
liquefy under the touch of the returning
sun, — that a mind of ordinary intelligence

which enjoys such a privilege will acknowledge the desirability and the practicability of fencing off very much larger portions of space with transparent material, so that multitudes may be enabled to enjoy the benefit and the pleasure of a mild and equable winter temperature.

An article published in "The Atlantic Monthly" for March, 1873, entitled "Life Under Glass," attracted considerable attention throughout the Northern States from people who are not afraid of ideas merely because they are new. It was suggested to the author from influential quarters to extend the essay, and have it published in a book form. The author is conscious, that, even in its enlarged form, the essay is still an inadequate presentation of a subject so important (as he believes it to be) to the well-being of the Northern peoples.

LIFE UNDER GLASS.

CHAPTER I.

INTRODUCTORY.

It is one of the tritest of axioms, that custom, or repetition, will often reconcile us to the most afflicting events.

Let a desolating war break out in any country after a long interval of peace, and the first insignificant skirmish excites the public mind to the most intense degree, although the loss of life may be slight; but, let the war continue long enough, and the most sanguinary battles at length cease to excite in the contending peoples, except in individual cases, that thrill of horror which attended the breaking-out of the carnage.

As with wars and battles, so with diseases. There are some destructive maladies which cause an annual mortality far greater than the loss of life in any battle of modern times, yet which have become so common, so closely inwoven into the fibre of the race, as to seem as much a part of the fixed order of things as are the taxes, and quite as little to be avoided.

At the head of the list of such diseases stands consumption, unrivalled by any other malady of the North in the number and character of its victims. This scourge of the most enlightened of the earth's peoples, those who boast of their descent from the energetic and the progressive Aryan race, loves a shining mark. Its too often fatal shafts seem to seek out the bright and beautiful of earth's children. It neglects the very young and the very old, but gathers its annual harvest of tens of thousands out of those in the early or later prime of manhood and womanhood.

Fifteen thousand human beings are annually killed by tigers in India. The North-American shudders as he reads a statement indicative of so deplorable a state of affairs, and thanks his stars that his lines are cast in pleasanter places. A little reflection would show him that there is an enemy among us more destructive of valuable life than all the tigers of India, *plus* its venomous serpents.

Of all the deaths that occur in most Northern countries, consumption is responsible for nearly or quite one-fifth. Let us suppose that the woods and swamps of our land were jungles infested by royal Bengal tigers, which caused a yearly destruction of life equal to one-fifth of all the deaths.

How long would such a state of things be permitted to continue? The tigers we are happily free from; but their place is more than supplied by an insidious and fatal disease, which, more discriminating than the scourge of the Indian jungles, selects its prey from the very flower of society.

The indifference with which this great loss of valuable lives is regarded in the communities in which it occurs is not flattering to the intelligence of the age. If some fatal disorder like the *pleuropneumonia* threatens the domestic animals, there is directly an intense excitement. Remedies of various kinds are experimented with; the whole matter is thoroughly discussed in newspapers, in farmers' clubs, and in public meetings. Legislatures are even convened to enact repressive laws, and stamp out the disease before it is too late.

In the case, however, of a merely human

malady like consumption, which sweeps off men and women in lieu of cows and oxen having a pecuniary value, society is content to fold its hands after the fatalistic manner of the Moslem, while the mortality proceeds unchecked, as an inscrutable and irremediable dispensation of Providence.

> "The fault, dear Brutus, is not in our stars,
> But in ourselves."

Yet society is roused from its apathy if the cholera in its infrequent visits, or the yellow fever, or the small-pox, claims a few victims. These are unmistakably contagious diseases; and selfishness whispers to each individual that it may be his turn next. Directly there is a panic: all the means that sanitary science can suggest are energetically used by boards of health having unlimited powers to prevent the spread of the malady. All of this is very natural and proper. If it

were possible, likewise, to get up a wholesome panic in regard to consumption, which
is a greater evil than all the other maladies mentioned combined, something might
also be done to check its ravages.

Much has been written as to the causes of
pulmonary consumption, and different theo-
ries prevail as to its nature and origin. Its
proximate causes are undoubtedly manifold,
the chief of which are, hereditary tendency
or a scrofulous taint of the blood, a weaken-
ing of the system, an unwholesome diet, &c.
But, whatever may be the proximate or the
exceptional causes, it is evident that an ungenial and variable climate must bear the
chief *onus* of responsibility for its preva-
lence.*   A climate liable to sudden and

---

* Possibly there are some who will not admit that consumption has its origin chiefly in atmospheric causes. If
that is not the case, why is it that there are certain climates
where the disease seldom or never originates? Such is the

great falls of temperature at all seasons, or which is subject to long spells of raw, humid weather, is shown by statistics to be a congenial habitat of pulmonary consumption. On the other hand, climates more uniform in character, and having a dry atmosphere, are comparatively exempt from lung disease.*

case in Minnesota, which is largely peopled from New England, the emigrants including many consumptive families; yet the children grow up without the disease developing itself, though retaining the same habits, and modes of life, as before their emigration. As a *remedy* for the disease when developed, the climate of Minnesota has been greatly overestimated. It is stated by good authority, that not more than five per cent of those in the earlier stages of this disease are permanently benefited by removal to Minnesota. The land is high, and the air consequently dry and pure ; but the terrible severity of the winters is a great drawback.

* To trace the connection between the languages of different peoples and the climates they lived in would repay the investigations of even a Max Müller. It is undoubtedly possible to judge very nearly of the climate of any country from a mere vocabulary of the words in

From the peculiarities of climate of New England and some other portions of the Northern States, one would naturally expect to find consumption a prevalent disease in these regions; and such is the fact.

daily use by the inhabitants. Thus the climate of England is characterized by a very large proportion of dark, misty, rainy, and cloudy days. Somebody has compared it to looking up a chimney when the day is fair, and to looking *down* the chimney when it is unpleasant. The words in the English language descriptive of foul weather far outnumber those used to describe fair weather. For example, take the adjectives commencing with the letter D, which are in common use by the English people when talking of the weather. A spell of foul weather might be described as being dark, damp, drizzly, dreary, dismal, dirty, dull, dripping, doleful, drowsy, dumpish, dubious, distressing, deused, dreadful, detestable, dangerous; and, if the colloquist were addicted to profanity, several more emphatic adjectives beginning with D might be used. To describe a fine day, using only words commencing with the letter named, the choice would be restricted to three or four dry,—delightful, delicious, and perhaps delectable, though the latter word is not in general use.

Whatever else we may have to be proud of, our climate is not a subject for unmixed admiration. It is well known to be a climate of extremes, — what the naturalist Buffon called an excessive climate, — extremes not only of heat and cold, but of wetness and dryness. The great range of the thermometer, in some years more than a hundred and twenty degrees of Fahrenheit's scale, is rivalled by the fluctuations of the hygrometer. Pluvial floods that would not discredit the rainy season of the tropics — times when it seems easy to believe in the theory of Leibnitz, that the universe is in flux — are followed or preceded by drouths worthy of the red sands of the Colorado desert, — drouths sharp and long enough to cause all organized life, animate and inanimate, to thirst for a little of that moisture, which, in its excess, was so great a discomfort to man and beast. It was such a drouth that prepared the way for the terrible *dies iræ* of the burning of Chicago and the Michigan lumber region.

At all seasons the temperature of these regions is liable to sudden and great alternations from warmth to cold, and the reverse. For instance, on the 29th of January, 1873, the temperature at the writer's residence, in Southern Massachusetts, fell fifty degrees in seven hours, or from thirty degrees above zero to twenty below that point. The next day the mercury rose seventy degrees in five hours. Such severe changes are destructive to vegetable as well as to animal organizations. In the winter of 1871–72 the combined cold and drouth were so intense as to destroy hardy evergreens over a large extent of the country. The following summer was intensely hot. More than three hundred fatal cases of thermic fever, or sunstroke, were reported in New-York City alone.

The mortality in that city, during the terribly hot week ending with July 6, was three times as great as the average.*

* A valuable and interesting feature of the ninth-census reports are the "charts of mortality" in the second volume. These charts show in distinct colors the relative mortality from various maladies in different sections of the country. The first shows the mortality from consumption. The highest average of deaths from this disease, over two thousand in ten thousand, is indicated by dark blue. This color covers the greater part of the New-England States, and also appears in Northwestern New York, in Eastern New Jersey, around the head waters of the Ohio River, in South-eastern Indiana, and Northern Kentucky. The next highest average, . fourteen hundred to two thousand deaths in ten thousand, is shown by a lighter shade of color. This covers a large part of Michigan, Southern Wisconsin, and parts of Illinois, Iowa, Ohio, Tennessee, and West Virginia. The greater portion of the Middle States are covered by it, excepting Pennsylvania. The only spots in the country east of the Rocky Mountains with a perfectly white record, with less than five hundred and fifty deaths in ten thousand, are the northern parts of Minnesota, Wisconsin, and Michigan, and small areas in Virginia, North Carolina, Georgia, and Southern Florida.

It would be easy to fill a large volume
with evidence showing the excessive and
variable character of our climate; but the
labor would be superfluous. Every resident
of the regions under consideration is
thoroughly sensible of the fact from unpleasant, personal experience. It is not
strange, that, in a climate with such peculiarities, *phthisis pulmonalis* is responsible for
one death in every five. A locality in which
a harsh, changeable winter and spring are
followed by a summer of debilitating heat
. is the most unfavorable that could be devised
for those afflicted with phthisis, or who have
a tendency towards it through hereditary
descent. Hence many, who have the means
and can bear the journey, take flight in the
autumn with the birds of passage to more
friendly climes, — to Florida, to Georgia, and
the Carolinas, to the Bahamas, and even to
Southern California. It has been denied by

some physicians of eminence — among them
the noted Dr. Ramadge of London — that a
winter residence at the South is of any
benefit to people having tuberculous diseases
of the lungs. There is evidence, however,
that many cures of the earlier stages of the
malady have occurred owing to removal to a
milder latitude. The number of such cures
would, no doubt, be much larger than it is
if any southern climate could be found
where the conditions were absolutely perfect for effecting a cure. Italy, Southern
France, and Spain, were formerly popular
winter resorts for those afflicted with this
disease; but experience has long since shown
that the shores and islands of the Mediterranean Sea are far from being favorable
localities for the cure or amelioration of this
disease. The average winter temperature is
mild compared with that of more northern
lands; but sudden changes frequently occur.

Cold, cutting winds, like the *mistral* of Southern France and the *tramontana* of the west coast of Italy, alternate with the hot and debilitating sirocco, "Auster's sultry blast." The spring months are especially trying to persons with weak lungs, from the keen, easterly winds which often prevail at that season. The best resort for consumptives across the ocean, excepting some parts of Syria, is undoubtedly the Island of Madeira. Yet the climate of this favored island is not perfection. "The spring at Madeira," says Sir James Clark in his work on the sanitive influence of climate, "as at every other place, is the most trying season for the invalid, and will require, even there, a corresponding degree of caution on his part." Notwithstanding this drawback, he considered a residence in Madeira, during the cold season, to be decidedly beneficial in the earlier stages of lung disease.

The winter climates of Florida, of Southern California, and of some other places in the austral regions of our country, with the summer climate of the Minnesota watershed, are far superior, in a sanitive point of view, to the shores and islands of the Mediterranean. Florida especially, though sometimes subject to rough " northers," has a winter temperature of great mildness. That of San Diego, in Southern California, is said to be nearly or quite equal to it in this respect.

But, whatever may be the advantages of these and other distant resorts for the sick, it is obvious that they can be made available to only a small portion of the large numbers of persons who need a mild, dry, and equable atmosphere as a primal condition of cure. Possibly five per cent of this class are able to bear the expense and endure the fatigues of the long journey required. What is to be done with the remaining ninety-five per

cent ? Must they give over all hope of recovery, and hasten the sad *finale* by yielding
to the depressing influence of a cheerless
gloom? At present there is only one re-
source by which they can avoid, to some extent, the trying changes of temperature, and
the cold, raw spells of weather incident to
the winters and springs of our northern
clime; and that is, to keep indoors as much as
possible. The remedy, it is needless to say,
is almost as bad as the evil sought to be
avoided. How can invalids regain health
who have to breathe for days and weeks at
a time the close air of a sitting-room,
poisoned by stove or furnace, and filled with
the irritating dust from carpets and clothing?
It is only the natural sequence of cause and
effect that the enfeebled vitality of multitudes succumbs under such unfavorable conditions. Is there no remedy for this state
of things? Cannot an artificial climate be

provided for these stricken ones, which shall
furnish a breathing medium, dry, pure, agreeable in temperature, and of a nearly uniform
degree of warmth? — in brief, a climate far
surpassing, in its sanitive influence, any
natural climate on the globe? The writer
believes that this is entirely possible; and
he is not without the hope of imparting
to others some portion of his own well-
grounded faith.

The question has often been mooted,
whether man has the power to influence or
change, in any degree, the general climate of
the regions he occupies. Those conversant
with the facts bearing upon the question can
have but one opinion to give in the premises;
which is, that, within certain narrow limits,
the climate of any country can be modified
by human agency. The temperature of large
districts in England has been raised appre-
ciably by the artificial drainage of the soil,

which was primarily cold through excessive wetness. After drainage, the dry, cultivated soil, being more easily warmed by the sun, absorbs heat to a great depth, and imparts it again to the air above by radiation.

In the neighborhood of Salt Lake, in Utah, cultivation of the soil and tree-planting have noticeably improved the climate. Where there were formerly frosts in every month of the year, frost is now unknown during the growing season. The climate has also lost much of its former aridity. Rains are said to be much more frequent; and the level of Salt Lake is constantly rising, threatening ultimate overflow of its banks. Possibly, in time, it will become a fresh-water lake. The wonderful patience and industry of the illiterate Mormons have changed the region they occupy from a cold, arid desert into a verdant garden, where all the fruits and cereals of the temperate zone grow in perfection.

Similar changes, though perhaps less marked, have attended the cultivation of the soil in other seemingly unpromising localities. Even under the proverbially rainless sky of Egypt, it is said, some indications of a moister state of affairs have followed the extensive planting of forest-trees under the orders of the Khedive.

Such instances as these show that meteorological phenomena can be sensibly modified by man's agency. But they do not prove that it will ever be possible to effect any radical change in the earth-climates. No amount of drainage or tree-planting would ever give to New England the atmospheric temperature of the Gulf States. Such a radical *bouleversement* as that, if it ever takes place, will be due solely to the operation of those occult forces, — whether cosmical or telluric in their nature is yet an open question, but, in either case, inconceivably slow,

as the mills of the gods, — which, during the
existence of our weather-worn planet, have
more than once revolutionized its climates,
and which may even now be inaugurating
the cycle that shall, perhaps, hundreds of
centuries hence, restore to these northern
lands the tropic temperature and vegetation
they possessed in the carboniferous age.
Such a contingency as that, if it were a near
one, would not be entirely agreeable to con-
template ; but it is altogether too remote to
excite any deep interest in the world of to-
day. What the present generation of men
is, or ought to be, interested in, is to use the
means it undoubtedly possesses to counteract
the ill effects of our present climate upon the
systems of those who are unable to bear its
severities of temperature.

It is obvious that no efforts of man will
ever enable him to control, in any degree,
the vast atmospheric waves which sweep

over land and sea with alternate floods of warmth and cold, — now soothing us into Elysian dreams with the soft accent, the *spiritus lenis,* of the sweet south; now scourging and pinching us with the *spiritus asper* of the icy north. Yet these great and often sudden changes of temperature can be rendered innocuous to the most susceptible invalids by creating isolated climates of considerable extent, and of any desirable temperature and hygrometric condition of atmosphere. The material which chiefly enables this to be done is abundant and comparatively cheap, and, were it not so common, would excite perennial wonder and admiration.

.There is no transformation in art or nature more wonderful than the conversion of substances so opaque, and so earthy in nature, as are sand and alkali, into such a material as glass, — a material having scarcely one property characteristic of its components; which almost rivals the diamond in hardness, brilliancy, and transparency; which can be blown, moulded, and cut into myriad shapes of use and beauty; which enables mankind to have light, warm, and cheerful homes; which furnishes the means of exploring the distant abysses of the stellar spaces, and of revealing a new world in the dust of earth and air, and by the aid of which it becomes easily possible to grow the flowers and fruits of the tropics under the cold skies of the north. There is no other substance of man's invention that approaches glass in its importance to the well-being of the race in high latitudes. Without its beneficent aid, large portions of the earth's surface that are now peopled by thriving and mentally-advanced communities would be uninhabitable except by semi-savages. Indispensable as it already is, its use will undoubtedly be greatly extended in the near future. Owing to its property of allowing the transmitted heat-rays of the sun to pass through its substance without material hinderance, while it prevents the free escape of the heat thus imprisoned into space, one can bottle up sunshine, as it were, in one's grapery or conservatory. If a man puts a glass roof over his garden, it is equivalent to a removal fifteen or twenty degrees nearer to the equator.

The writer has had some experience as an amateur cultivator of our native grapes. Some years ago he had a mild attack of the grape-fever, induced by the glowing, descriptive catalogues of certain noted vine propagators in the valley of the Hudson. A spot of mellow garden-soil was prepared by trenching and enriching ; a close board fence was built around the north and west sides for a wind-screen ; and in this dry, rich, sheltered spot, several hundred vines were planted, comprising all the varieties then in vogue. For two or three years the vines grew well, and some first premiums for the fruit were taken at the county fairs. But a succession of unfavorable seasons ensued. Heavy floods of cold rain in July and August, followed by scorching sunshine and sultry air, brought mildew and sun-scald in their train. Sulphur, said to be a specific for the *oidium*, was freely applied through the nozzle of an old bellows, but with scarcely any effect. The diseased leaves dropped from the vines in untimely showers before the end of summer, leaving the unripened fruit to shrivel and perish. Deprived of the leaves, the wood of the new shoots also remained unripened, and the buds for the next year's growth undeveloped. As a consequence, all the vines except a few of the hardier sorts were permanently enfeebled.

The impending failure of the grape cul-

ture in the open air being foreseen, it was
decided to transfer the field of operations to
an artificial climate. A grape-house of
modest dimensions was built, and sixteen
vines of the best foreign varieties were
planted with their roots in the border outside. No more pains were taken with the
border than had been bestowed on the soil
in which the out-door vines had been
planted. The house-vines were not set
until June, and not much growth was
expected of the feeble-looking little things
for that season. They soon, however, sent
up shoots which grew with wonderful rapidity and *vim* in the genial air of the grapery.
They seemed to be almost conscious of their
good fortune in being placed in such favorable conditions for growth, sheltered from
rough winds, from cold, and from storms. It
was a pleasure to watch their daily develop-
ment of healthy, stocky wood, and of richly-

colored, beautifully-shaped, veined, polished, crenulated leaves. Some of the canes attained a length of more than twenty feet, with a diameter at the base of three-fourths of an inch, before they were checked by the cold of autumn. The health and vigor of the vines still continue undiminished, and they annually bear many clusters of fine fruit.

Glass graperies are not uncommon nowadays; and the writer's experience in this interesting department of horticulture would hardly be worth relating except as an illustration of the superiority of an artificial climate over a harsh natural one for the cultivation of semi-hardy and tender vines and plants. Foreign grapes, varieties of the *Vitis vinifera*, it is well known, will not succeed in the Atlantic States, even when planted several degrees farther south than their original homes in Europe and Asia.

If one has a greenhouse in his garden, he is, to a certain extent, independent of the changing seasons and of inclement skies. In winter, when the sun shines, he can enjoy a summer-like temperature under the protecting glass; and, by adding artificial heat, he can surround himself with greenery and bloom. Even when the sun is veiled by thick clouds, its heat-rays penetrate through the vapors of the upper air, so that the temperature of the air within the greenhouse is very perceptibly raised. To the susceptible invalid in winter, weary of the prison-life of a sitting-room, perhaps ill lighted and ill ventilated, a greenhouse on a sunny day seems a delightful change. The gales of January or of March may be roughly scourging the world outside ; but under the glass roof the air is quiet and genial like a winter's morning in the Antilles. The enfeebled system absorbs the magnetic, vitalizing

rays in every pore. Our style of domestic
architecture will not, as a general rule, allow
of having, like the old Romans, *solaria* on the
house-tops. Our *solaria* for the enjoyment
of sun-baths must be on a less-elevated situ-
ation; and, for a large class of invalids in
winter, must be under glass.

It will not be questioned that tender plants
which have been enfeebled by exposure to a
harsh atmosphere can generally be restored
to vigor by simply placing them under the
shelter of a glass roof. Considering the es-
sential solidarity, and oneness of origin, of
the two organized kingdoms of Nature, their
close similarity in their lowest forms, so that
scientific men yet disagree as to the point
where the vegetable kingdom ends and the
animal kingdom begins, it would seem no
more than a just inference, that, in their more
developed forms, they must still have many
attributes or properties in common. Is it

not, then, reasonable to suppose that the
treatment required by delicate, exotic plants,
which have become diseased in an ungenial
climate, is, with some modifications, the
treatment proper for animals and for human
beings enfeebled by the same cause? For
man also, under these inclement skies, is not
yet acclimated, but may be considered an
exotic from the warmer regions of the planet
which undoubtedly gave him birth.

A few years since, a gentleman of Philadelphia — Gen. A. J. Pleasonton — instituted
some experiments with animals placed under
glass, or in glass-covered pens, in winter.
He found that healthy animals, such as
young pigs, grew more rapidly than in ordinary pens; while a sick calf was speedily restored to full health, and made a surprising
growth. The glass used in the pens was
half blue or violet colored, and half common
or colorless. The experimenter attributed

the beneficial effect to the blue glass alone, from the previous wonderful effect it had seemed to produce upon the growth and fruiting of some vines in a grapery in which it had been used. Yet it is more than probable, that, if all the glass in the pens had been colorless, the same, or perhaps even a superior effect would have been produced upon the animals. The time of year being winter, they had the full benefit of the sun's light and warmth, with no exposure to cold winds and storms.

These experiments * were not thorough enough to be conclusive upon any point; and

* Some experiments of the writer's with fowls having the roup and other disorders showed conclusively that recovery was much more certain and rapid when the sick birds were placed under the shelter of a greenhouse than when confined in an ordinary pen, or allowed to roam at large.

The claim of the Philadelphia gentleman in regard to the efficacy of the violet rays in promoting vegetable and they certainly do not warrant the conclusion that the beneficial effect was owing solely to that portion of the glass which was colored blue. As far as they go, they simply coincide with the inference deducible from a common-sense view of the subject; which is, that domestic animals, whether sick or well, will thrive better when protected from the storms and cold of winter, if, at the same time, they can have the benefit of the sun's light and warmth, than they will in pens of the ordinary construction.

animal growth is not sustained by later experimenters, such as Professor Pfeiffer of Marburg; and by Selim and Placentim. The experiments of these gentlemen show that the yellow rays are more promotive of the evolution of carbonic acid in animals, and its absorption in plants, than any other color in the spectrum, or than white light; the violet rays having the least power in these respects, excepting the red rays in the case of animals. The absorption of carbonic acid by plants, and its evolution by animals, are prime essentials to the growth and health of each.

From the great similarity of the human organization to that of the other mammalia lower in the scale of being, it is reasonable to infer that a like beneficial effect would follow if invalids with certain diseases having their origin in vicissitudes of weather, among them consumption, were placed, during the inclement season, in an environment permitting the fullest entrance to the sun's warmth and light, while, at the same time, the unfavorable influences of a rough atmosphere were excluded. Such an experiment — if that can be called an experiment which appears to the reflective intellect an absolute certainty — can only be satisfactorily tried on a very large scale. If properly carried out, it would call into exercise all the resources that modern science, aided by a lavish use of capital, can command; and the results would undoubtedly be commensurate with the means employed. In the succeeding pages, it is hoped to demonstrate, beyond reasonable cavil, that such an investment of money would be profitable to capitalists, as well as greatly beneficial to invalids.

CHAPTER II.

MANY millions of treasure directed by the best practical science have been expended in costly glass houses for the protection of rare plants, or for the growing of non-hardy fruits; but neither thought nor money has been given to furnish adequate winter shelters for the myriads of tender human plants whose physical systems are too weak to endure the rough weather of a harsh and capricious clime. Yet there are indications that the time is not very distant when this defect in our civilization will be remedied. The means for remedying it can be found only in a system of winter gardens, which,

40

as will be shown in the sequel, shall also be summer gardens during the warm season, on a scale of magnitude startling to · timid minds, and in which the chief materials of construction shall be iron and glass.

The London Crystal Palace of 1851 demonstrated the wonderful adaptability of iron and glass in combination for the construction of large edifices.   The history of this famous building is still fresh in the minds of this generation of readers.   Most of them can recall how marvellously rapid was its construction; its exceeding cheapness, costing less in proportion to its size than an ordinary barn; and how, having admirably answered its purpose as an exhibition-building, and attracted the wondering admiration of people of all nations, it was taken down, and reconstructed, with some modifications, at Sydenham, as a permanent pleasure-resort for the people of London and vicinity.   For beauty,

convenience, and cheapness of construction,
combined, the Crystal Palace for the World's
Fair of 1851 stands unrivalled among the ex-
hibition-buildings that have adorned the capi-
tals of the world. The sequence of events
that led to the realization of this unique and
splendid structure may be unfamiliar to some
of the readers of these pages.

In the year 1831, a gigantic species of
water-lily, with floating leaves many feet in
circumference, each of them capable of supporting, without sinking, the weight of a
half-grown boy, and with flowers of corresponding dimensions, was discovered by
Schomberg, the botanist, in the Berbice, —
one of the sluggish rivers of Demerara. Seeds
of this immense water-plant were sent to
England, and fell into the hands of Joseph
Paxton, then horticultural manager of Chatsworth, the princely seat of the Duke of Devonshire. A single plant raised from these

seeds was treated with the most careful attention to its supposed needs. It was placed
in a soil of peat and burnt loam; the temperature of the house was nicely regulated;
the water in its tank was artificially warmed,
and its surface was gently agitated by ma-
chinery in imitation of the ripples of a river.
Under such extra care, the plant expanded
with wonderful rapidity: it soon outgrew
the accommodations provided for it, and it
became indispensable to speedily enlarge its
habitation. Being a man fertile in resources,
and having *carte blanche* from the wealthy
and liberal-minded duke as to expense, Mr.
Paxton accomplished this in a few weeks;
and the result was a beautiful structure of
iron and glass, in which the *Victoria Regia*
could expand and blossom as freely as in its
native stream. The conservatory thus has-
tily erected was the germ of the Crystal
Palace, for the plans of which Joseph Paxton
was made Sir Joseph.

Thus, if the *Victoria Regia* had not been discovered; if seeds of the plant had not been sent to England; if these seeds had not found their way to Chatsworth, or had failed to germinate; and if a less ingenious person than Paxton had taken them in charge, — then the marvellous glass palace of the World's Fair would not have been built, but, instead, Hyde Park would have been lumbered with the ungainly pile of bricks and mortar decided upon at first by the building committee: so that, if any link in the chain of seemingly incongruous and unimportant circumstances had been lacking, the present age would probably have seen no adequate exemplar of the grand results that may be effected when iron and glass are used as the chief materials in the building of large edifices.

The conservatory at Chatsworth, though, of course, it could not compare in size with the World's-Fair building, was yet a splendid specimen of a plant-house, unequalled at that day in size and appointments. Two acres of glass panes were required in its construction, and it contained several distinct climates to suit the needs of plants from different zones. Some idea of its size and of the ultra regal splendors of the ducal palace may be formed from the fact, that, while Queen Victoria was once visiting at Chatsworth, she entered the conservatory one evening with the duke in a carriage-and-four, while the beautiful structure blazed and glittered in her honor with the light of fourteen thousand burners. Enchanted with the brilliant scene, the young queen turned to her host, and exclaimed, "Devonshire, you beat me!" If such marvellous glass structures have already been built for conservatories and for exhibition-purposes, what may not be expected in the future, when the

same material shall be used for the greatly
more important purpose of constructing sanitary resorts for the prevention and cure of
some of the worst maladies that afflict and
almost decimate the Northern peoples? For
it will yet be generally acknowledged, that
the great loss of valuable life, due to the ill
effects of extreme and changeable climates,
can be very largely reduced through the
beneficent agency of the material called
glass.

In the present stage of our civilization,
human life, judging from the general apathy
in regard to its needless waste, is considered
to be of less importance than the lives of
brutes. Little heed is paid to the hard logic
of statistics, which shows that more than one-
third of the human race, or at least of that
portion called civilized, dies before reaching
the age of five years. " Of all the coffins
that are made in London," wrote Charles

Dickens, "more than one in three is made
for a little child." Add to this slaughter of
the innocents the needless mortality at more
advanced stages of life, and the aggregate
becomes so enormous as to make it seem a
marvel that even the present slow increase
in the population of Christendom is main-
tained. In a new country like our own,
with an immense area of unoccupied terri-
tory, the subject assumes an importance,
which, from the merely utilitarian point of
view, it does not possess in the more dense-
ly-peopled countries of the globe. Until
our vast domain is filled with happy homes,
we shall continue to welcome the throngs of
emigrants that seek our hospitable shores
from the countries over the ocean. They are
considered, and rightly so, to be a most important element in the enhancement of the
wealth and power of the nation. Yet while
the peasants of Europe are encouraged to

emigrate, even by the lure of free home-
steads, very little heed is paid to the needless waste of life that is constantly taking
place in our native-born population.

If each raw Irishman or German that
lands from the emigrant-ships, with no other
wealth than the ability and will to work for
a living, is worth, as has been estimated, not
less than a thousand dollars to the country,
then certainly a native American, in the
same circumstances, is worth at least as
much. If he is threatened or attacked by
maladies liable to a fatal termination, and his
health is restored by the use of proper
means, then there is a thousand dollars, or
its equivalent, saved to the wealth of the
nation. Is it not about time to supplement
the Poor Richard maxim, that "a penny
saved is as good as a penny earned," by the
nobler one, that a life saved is as good as a
life imported? Often, in fact, it might be a

great deal better, considering the character of some of the human raw material that comes to us from the Old World. The utilitarian, the dollars-and-cents view of the matter, is not, of course, the highest one that might be taken ; but it is the one most likely to be heeded in an age when the first question asked in relation to any undertaking is, " Will it pay ? "

The site of a sanitarium, it is obvious, should be on high land, to secure the prime essentials of pure air and thorough drainage of the soil. A level-topped hill of sufficient area, with a gravelly subsoil, would be one of the best possible locations. The establishment, as has already been said, should be on an extensive scale. Only on a generous scale of expenditure could satisfactory results be produced, either remedially or financially. The form of the main edifice, whether the ground-plan is a circle, a square,

4

a parallelogram, a Greek cross, or other
figure, is not, for purposes of illustration,
very material; though an approach to the
compact form of the square or circle would
be much preferable to a long, narrow, cen-
tipede-shaped structure like the building for
the Vienna Exposition. The chief materials
of construction would be iron and glass.
These materials necessitate a light, airy,
graceful style of architecture. As the circle
is the simplest and most economical of materials for enclosing a given space of any
geometric figure, we will suppose the
ground-plan of the edifice to be round;
though the mechanical difficulties of erect-
ing a circular building would be somewhat
greater than if the shape were rectangular.
The diameter of the edifice should then be
at least fifteen hundred feet. The walls of a
circular building of this size would enclose
an area of a little over forty acres, — not far

from the size of Boston Common, exclusive
of the Public Garden. There may be some
who will consider the building of edifices of
such immense size entirely impracticable;
but, to the resources of modern engineering
and mechanical skill, there is, practically, no
limit in this direction, except the amount of
capital at command. Buildings of more
than half the size proposed have already
been erected. The London - Exhibition
building of 1862 covered an area, with the
picture-gallery and annexes, of twenty-four
and a half acres. A forty-acre building is
quite within the limits of the practicable.

The walls of the edifice would be forty or
fifty feet high, and would be supported by
iron columns of the proper dimensions,
bedded in concrete at the base, and likewise
by a cordon of ornamental iron towers
eighteen or twenty feet in diameter, rising
above the walls. The vast expanse of the

glass roof would be supported by regular
rows of lofty iron columns, upholding the
symmetrical system of arches, girders, and
rafters overhead. The roof would rise at a
gentle, regular pitch from the walls towards
the centre, where it would be nearly a hundred feet from the ground below. The
glass panes covering it would be no less than
one-third of an inch in thickness, and four
feet long. The sashes they would be fitted
into would be arranged on the well-known
ridge-and-furrow principle, with alternate
angular depressions and elevations, the lower
angles forming gutters to carry the rain-
water into the hollow columns, whence it
would flow into the system of drain-pipes
and sewers. In the centre of the edifice
would rise a dome of lofty proportions,
artistic and graceful in its outlines and
adornments. This dome would be furnished
with ample ventilators worked from below.

The steam-boilers for warming the edifice in the absence or inadequacy of the sun's rays would be in the lower stories of the towers outside the walls ; which would thus serve a double use, besides being a pleasing architectural feature of the building. Pure air would constantly pass into the interior of the garden through a sufficient number of apertures near each tower in the lower portion of the walls. This fresh, cold air would be warmed on its passage into the building by passing through net-works or coils of hot steam-pipes. There would thus be an abundant and constant, though gentle, flow of pure, warm air from all points of the base of the garden towards the centre ; where, in accordance with the familiar law governing heated atmosphere, it would rise and flow out through the ventilators in the dome. Such a constant influx at the base and efflux at the top of the building would insure the greatest purity of the air within. It would have none of the oppressiveness often experienced in a common, ill-ventilated conservatory, but would be in the highest degree agreeable and healthful. The means of ventilation would be under such easy control as to enable those in charge of that department to maintain a nearly uniform temperature, whatever might be the fluctuations of temperature outside.

The æsthetic details, the adornment of the grounds inside, will now be briefly considered. Although the constructive details of form, size, materials, &c., have to be first considered, yet the adornment of the grounds within the walls, and the furnishing of ample means to gratify the various mental needs of a large community of people of diverse tastes, is a question of the highest importance to the success of the undertaking. In this, as in all other departments, considerations of expense should not be allowed to hinder the grand result intended to be accomplished; which would be the creation of an arboreal and floral Eden, where the most consummate art would be so concealed as to seem only nature under superior conditions.

The area of forty acres would require for its laying-out and embellishment the careful thought of the most accomplished landscape-gardeners. All the attractions that un-stinted means can command for the adorn-ment of pleasure-grounds would be here called into requisition. Broad, winding, gravelled avenues and serpentine paths would lead among rock-work, shrubbery, clumps of ornamental trees, trim lawns, and parterres of flowers; over ravines spanned by graceful bridges; around miniature lakes and fountains; and by the side of grassy banks, where the winter sunbeams would

linger as warmly as if it were June. There
would be swings, and archery and croquet
grounds; there would be aviaries of birds
from all climes; there would be a large and
well-stocked aquarium, and a zoölogical
garden containing specimens of such animals
as are especially noted for interesting tricks
and manners, or for beauty of form and
coloring. Music would lend its subtle
charm; birds of song would flit among the
tree-tops or in the shrubbery; the bell-like
notes of the hermit-thrush, the haunting
sweetness of the veery's song, the gushing
joy of the bobolink, the cheerful refrain of
the song-sparrow, the flute-like call of the
oriole, the robin's clear madrigal, and the
blue-bird's warble, would call up reminis-
cences of the bright days of early summer,
though winter might still desolate the world
outside. At occasional times, the music of
a large and perfectly-trained orchestra would

set the warm air of the garden pulsing with the lively, the martial, or the grand religious strains of the great tone-masters of the world.

The hundreds of lofty iron columns required to support the crystal roof and the vast dome would be made to harmonize with the scene by having their formal outlines concealed under a covering of rapidly-growing, climbing vines. They would thus resemble rather the tall trunks of a symmetrically-planted grove of palm-trees, wreathed with tropic climbers, than the hard, unattractive supports of the edifice. Scattered everywhere, singly and in social groups, in sunny nooks and cosey corners, would be found an abundance of the most inviting seats and lounges. Seated or reclining in these after the needful exercise of the day, the invalid visitors could pass the time in any rational way to which they felt inclined,

in some light, agreeable work, in reading, in
conversation, in games of chance or skill, or
in observing the animated, enchanting scene
around, while listening to the inspiriting
strains from the orchestra, or the softer
melody of the singing-birds.

Within the transparent walls of a palace-
garden of the size designated, ten thousand
or more visitors would find ample room for
exercise and recreation, an atmosphere pure
and agreeable in temperature, plenty of
opportunities for taking sun-baths, agreeable
society, and countless objects of interest to
occupy and interest their minds. The un-
healthy mental habit, so common with invalids, of an introverted, anxious study of
their own symptoms, would give place to a
lively and healthy interest in their novel and
delightful surroundings. They would almost
forget that they *were* invalids amid the manifold attractions on every hand. With none

of the unfavorable winter conditions of
ordinary house-life to contend with, the
recuperative powers of the human organization, the *vis medicatrix naturæ*, aided by the
pure, mild air, the regular exercise, the
genial sunshine, albeit of midwinter, and
the more healthy mental status, would, in a
large majority of cases of *phthisis pulmonalis*
in its earlier stages, and in some other
diseases, soon show the happiest results ;
the irritated, tuberculous lungs would grad-
ually heal; the wearing cough would sub-
side ; the pains of the rheumatic would yield
to the sanitive influences of the place, and
retire into some unknown limbo, where are
gathered those undesirable things, which,
when happily lost, are never sought for
nor regretted;* strength would return to
weakened frames, roundness to wasted limbs,
and happiness to clouded minds.

* *Damnum absque injuria,* as the lawyers say.

It is admitted that these are rose-colored pictures; but they are in no degree over-drawn. They fall short of what will be realized when even a small fraction of the outlay, the thought, and the time, that are now given to the destruction of human·life, shall be given to its preservation. So long was it taught in the schools of medicine, *ex cathe-drâ*, that consumptive disease was, from its nature, wholly incurable, that the idea still influences a large portion of the medical faculty, as well as of the public at large.*

* The science of medicine in its present state has received the hardest blows from its own professors. Here is what Dr. John Mason Good says in the premises: "The science of medicine is a barbarous jargon; and the effects of our medicines on the system are in the highest degree uncertain; except, indeed, that they have destroyed more lives than war, pestilence, and famine combined." Similar testimony is given by other distinguished physicians, among them our own Dr. Holmes, who thinks, that, if the greater part of the *materia medica* were thrown into the sea, it would be better for mankind, and worse for the fishes.

No doubt it *is* incurable by drugs, whether in heroic or infinitesimal doses: but give Mother Nature a fair chance, furnish the proper conditions, and she will often perform marvellous cures. These conditions, it is obvious to the candid, receptive mind, would exist in perfection in a winter-garden like the one described, — not wholly described, however; for some important details have not been referred to.

The large numbers of people that would congregate at an establishment like the one under consideration would need the best of boarding accommodations close at hand, so that no exposure to cold, stormy weather need occur in going to or from the hotels. The plan includes a broad, glass-enclosed street, extending entirely around the central edifice, at the distance of, perhaps, three hundred feet from its walls. This circular boulevard would be at least seventy-five feet

wide, and would have an ample carriage-
way in the centre, paved with wood or
asphalt, and at the sides roomy, level walks,
separated from the carriage-way by iron
railings covered with flowering vines. The
walls of this boulevard need not be more
than one-third as high as the garden-walls,
and the roof would be arched. It would be
warmed and ventilated like the garden, with
which it would be connected by glass passage-ways. Here would be the finest and
most unique of street-arcades, — a crystal-
covered, circular Broadway, or Boulevard des
Italiens, more than a mile in circuit, adapted
for drives, for horseback-riding, or for prom-
enading, and available for use by the most
susceptible invalids in all weathers. Let the
storms of winter rage as fiercely as they
might out of doors : they would never disturb
the serene quiet and warmth of this circling
arcade. On its outer circumference would

be the spacious hotels for the visitors, connected with it by short, enclosed passageways. The hotels, twenty or thereabouts in number, would be managed by thoroughly competent and trustworthy persons, who would see that all reasonable wants of their guests were provided for. The food furnished would be nutritious and wholesome. The warmth and ventilation of the apartments would correspond with the rest of the establishment. In the persons of the landlords would be united the characters of the genial, considerate host, and of the intelligent physician. Like the other officials of the place, they would be picked men.

Between the garden-walls and the encircling boulevard there would be a ring of ground open to the outer air, three hundred feet wide, and extending around the central building. The radiating passage-ways from the garden to the boulevard, eight in number,

would divide this ring of ground into an
equal number of distinct plats, each several
acres in extent. These open-air gardens
would be tastefully laid out with walks,
grass-plots, evergreen-trees, and shrubbery:
surrounded on all sides by high walls, they
would be sheltered from rude winds, and
would afford fine opportunities for exercise
on mild, sunny days. They would also contain the spacious buildings required for gym-
nasia, libraries, chapels, galleries of art,
museums of natural history, theatres, bowl-
ing-alleys, and the like, needed for the exercise, amusement, or instruction, of the visitors.
All of these buildings would directly commu-
nicate with the garden, or with the arcades.
There would also be numerous shops, of
various kinds, along the boulevard. Shopping, that occupation so congenial to the
feminine mind, and not wholly devoid of
attractions to the rougher sex, could thus be
done in all weathers.

Excepting at meal times, and during the hours required for sleep, but little of the time of the visitors would be passed in the hotels. Even the evenings would be chiefly spent in the garden and the arcades, which, lighted by thousands of burners, or by shaded, electrical lights, would seem like some enchanting dream of romance. Foliage never appears to such beautiful advantage as under a strong artificial light: the shadows are deeper than under the more diffused light of day; so that the graceful spray and finely-cut leafage of rare trees and plants, in an illuminated pleasure-garden, are brought out with salient and picturesque distinctness. With the brilliant light, the sparkling fountains, the cheerful warmth, the beautiful foliage and flowers, the moving throngs of people, and the delicious music of the perfectly-trained orchestra, the garden-palace in the evening would be another " Pal-

ace of All Delights." For those who chose
to attend, there would also be the special
attractions of the theatres, concert-halls, lec-
ture-rooms, and other outside entertainments.
Amid such an endless variety of attractions,
there would be no opportunity for indulging
morbid fancies or forebodings. Cheerfulness
would take the place of despondency, or of
oft-recurring apprehensions for the future;
and thus the medicament of the healing
forces of Nature would have a fair field for
its restorative effects.

Works have been written to show the
powerful influence of the mind over the
body, either for good or ill; but this extraor-
dinary influence of mental conditions in its
sanitive aspect is not recognized, nor taken
advantage of, as it should be, by any system
of therapeutics yet established. Imagination is a more potent remedy in disease than
all the drugs in the pharmacopœia. Well-

authenticated instances show that it is sometimes capable of causing death, and its re-vitalizing power over its physical integument is equally well established. Napoleon's saying, that "imagination rules the world," is no less true of the little world, the microcosm of man's personality, than of the world of affairs. In our ideal establishment, the influence of the imagination would be taken advantage of in all possible ways to assist in overcoming disease.

In the foregoing necessarily imperfect description of what is desirable in a remedial establishment for large classes of invalids, there is nothing that cannot be easily realized, when the importance of the subject shall be impressed, as it should be, on the minds of philanthropists and capitalists. As a paying investment, such a winter-resort would undoubtedly surpass most of the popular stocks that command a premium on Wall or State

Street. Although the data are not all attainable on which to base an exact calculation of the cost of constructing and of operating a sanitarium on the scale indicated, yet it is possible to form an approximate estimate of these expenses, as well as of the probable returns or profits.

In an article in " The Atlantic Monthly " for March, 1873, and on which this essay is partly based, some calculations were entered into as to the amount of capital needed, and the cost of operating, as well as of the pecuniary returns that could be reasonably expected. The Crystal Palace of 1851 was taken as a basis of calculation. That edifice contained thirty-three million cubic feet of space ; and it cost at the rate of one penny and one-twelfth per cubic foot, or an aggregate of one hundred and fifty thousand pounds sterling. Our ideal sanitarium would contain about four times the cubic space of the

Crystal Palace, including garden, dome, towers, and arcades, in the calculation, but not the hotels and other outside buildings. At the same rate per foot as the London Palace cost, it would require the sum of six hundred thousand pounds, or about three millions of dollars, in its construction. Owing, however, to the much higher prices of labor and materials in this country than in England, and their general advance in both countries since 1851, our garden-palace would cost more than twice the amount named. To be on the liberal side, we will estimate its cost at eight millions of dollars of our currency. For the land and its grading, drainage, and ornamentation, and for the hotels and other needful structures, four millions more would most probably be sufficient. Twelve millions of dollars * would, therefore, be the

---

* This amount is about the estimated cost of tho Hoosac Tunnel, when completed.

estimated capital required. The interest on
this amount, at the liberal rate of eight per
cent, would be nine hundred and sixty thousand dollars a year. The working expenses
are somewhat more difficult to estimate. A
somewhat careful study of the subject leads
to the conclusion, that the running expenses,
including the cost of boarding ten thousand
visitors from the 1st of November to the
1st of June, would not exceed $2,500,000 a
year. Adding this amount to the interest on
the capital, we have the sum of $3,460,000
as the outgoes of each working year. To
meet these expenses would be the board-bills
of the visitors for the season.

The price of board at the hotels should be
placed as low as possible, so that people of
limited means could enjoy the benefits of the
sanitarium as well as the rich. Large numbers of people can be provided for at a much
lower rate *per capita* than small numbers

would cost. Two dollars a day will not seem an unreasonable price, when it is considered that it is much lower than the cost of living in first-class city hotels or at any popular Southern winter-resort, and that here all the inestimable advantages of the garden would be thrown in. At two dollars a day, the board-bills of ten thousand visitors for thirty weeks would foot up the very large sum of $4,200,000, or $740,000 more than the interest on the capital at eight per cent and the estimated working-expenses united. This surplus would certainly furnish a reserve-fund large enough to meet any unforeseen outlay that might occur.

Let no person of little faith suppose for a moment that there would be any lack of visitors at a winter-resort like the one under consideration, even if the *per diem* were twice the rates proposed. "All that a man hath will he give for his life." During the

time that would be required to complete such
an establishment, its great magnitude, its
novel plan, its unexampled attractions, and
its presumable sanitary advantages, would be
thoroughly canvassed, not only in our own
land, but all over that part of the globe to
which the newspaper-press has access.   It
would be advertised gratuitously among all
peoples, and invalids of every Northern
country would desire to avail themselves of
its undoubted benefits.   From the opening
day the hotels would be filled to their ca-
pacity with the weak-lunged, the rheumatic,
and the declining.   But the phthisical and
the rheumatic would not be the only classes
to whom the winter-garden would be a delightful haven of rest, where the constitution
could recover from the injurious effects of
exposure to a stormy world, or of an ignorant
disregard of hygienic laws.   Carlyle, in one
of his dyspeptic moods, — if, indeed, he have

moods of any other sort in these latter days,
— tells with grim humor of the time when
he first became aware that he possessed a
diabolical apparatus called a stomach. In a
land where pork in all its Protean forms,
and hot saleratus-bread, form the chief sta-
ples of diet, there are multitudes who have
arrived at the same disagreeable knowledge
in regard to their epigastric region that dis-
turbed the Chelsea sage. There is no doubt
that some forms of dyspepsia, for which a
mild winter climate is recommended, would
readily yield to the magnetic warmth, the
wholesome, well-cooked, nutritious food, and
the systematic exercise, furnished by the proposed sanitive establishment. And there
are still other classes to whom a residence
within its walls would insure speedy relief
from their maladies, and, finally, a permanent
cure: one such class, for instance, would
consist of the vast army of performers on

what Charles Dickens called the "great
American catarrh," although the instrument
is well known in England and other countries.

Who will have the hardihood to say that
the winter-garden would lack patronage,
when there are so many times ten thousand
people in the land, without including visitors
from other lands, who would anxiously watch
its progress to completion, and joyfully avail
themselves of its unexampled advantages?

There is, however, another view of the
subject, well worthy of having a volume
written in its elucidation, but which can
only be briefly touched upon here; and that
is the advantages of such a winter-resort for
well people.   Even if an acknowledged
invalid were never allowed to enter its gates,
a place with such marvellous and unique
charms as the winter-garden would possess
would be thronged for half the year by

people of wealth, culture, and fashion from all parts of the land. No city in America can at present offer such allurements to men and women of means — whether they were refined and intellectual, or sensuous and superficial — as would be concentrated within the limits of the establishment. New York's summer pride and joy, the beautiful Central Park, would appear bleak and barren under a wintry sky compared with the wealth of verdure and of bloom to be found under the forty acres of crystal forming the garden-roof. The *élite* of the great cities would flock to it, as in summer they seek the cool breezes of the mountains, or of Newport, Long Branch, and Cape May. Within its sheltered precinct they would escape all the multiplied discomforts of city streets in winter and spring, — the driving storms of rain and snow, the blustering north-west gales, the slippery sidewalks, the

almost endless mud and slush, and the blind-
ing clouds of dust, that make a Northern city
in winter, for all its social, artistic, and lite-
rary attractions, a most wretched place of
abode for sensitive people. The contrast
between the comfortless streets and squares
of the cities in winter, with the verdant
beauty and the quiet warmth of the winter-
garden during the same season, would be too
great to pass unheeded among the wealthy
residents of Northern towns. At the garden
they would not only escape all the discomforts named; but they would find, besides
warmth and greenery, all the means for
amusement or culture they left behind in the
storm-scourged city. Operas, concerts, theatres, lectures, libraries, galleries, museums,
— all of high excellence, — would occupy the
charmed hours. The broad, circling, arch-
roofed boulevard would be a more thronged
and fashionable drive on a mid-winter's day

than even the avenue at Newport on an August afternoon. Owners of fast trotters and of stylish turnouts would all be anxious to display their teams and themselves on such a novel and magnificent track before the assembled wealth and fashion of the land. Each ingredient that goes to make up the summer crowd at a great watering-place would find its counterpart here, — fortune-hunters of both sexes, mammas with marriageable and unmarriageable daughters, fast young men, and flirting, gay young women. All of these well-known types would muster in their usual force; but there would also be very many visitors of a different stamp, — people of refined manners and cultivated minds, whose combined influence would be felt in the tone of the place, and would not be without its beneficial effect on even the brainless fops and feminine votaries of millinery. Poets, artists, essayists, novelists, would find here endless suggestions and materials to work into poems, pictures, essays, and stories. Especially would the many educated, delicate, sensitive, *spirituelle* young and middle-aged women that are to be found in wealthy families, and whose poetic, impressible natures, and general frailness of organization, make them keenly alive to the discomforts caused by the atmospheric changes of winter, find an earthly paradise within the genial realm of glass, where winter and rough weather were obsolete terms.

It has already been shown, that, if any reliance whatever can be placed upon *à priori* estimates, such a winter gathering-place would be a profitable investment of capital; but an important source of income was not mentioned. A location would be chosen where land was comparatively cheap, and a tract of several thousand acres secured. Two or three hundred acres immediately

surrounding the establishment would be
reserved for an outside park. It would be
finely laid out, and ornamented with drives,
foot-paths, skating-ponds, groves of decidu-
ous and evergreen trees, trim hedge-rows,
and shrubbery. This outer park would be
the pleasure-resort of the visitors on fine,
sunny days. The remainder of the land
outside of the park would be surveyed and
mapped into lots, streets, and squares, to
meet the requirements of a large, prospective
population. People of all trades and occupa-
tions would be drawn towards the city of
glass, to supply the wants, real or fanciful, of
its thousands of inhabitants. Men of wealth
and taste would surround the park with
ornamental villas. A large and prosperous
town, ultimately to grow into a city, would
inevitably soon crystallize within sight of the
lofty dome of the garden-palace, — a Dome
of the Invalides of even more magnificent

proportions than the famous landmark of
strangers in the French capital. Building-
lots could not be otherwise than in brisk
demand in the vicinity of a magnet so
powerful. The income to the corporation
from this source alone would be very great.
Four large, distinct towns have sprung up
around the Sydenham Palace since its erection a few years ago. Another source of
income would be the rent of stores along the
circular boulevard, or the lease of land for
their erection by other parties.

Thus far the sanitarium, or the garden-
palace, has been considered only in its winter aspect; but, paradoxical and improbable
as it may seem, it can be demonstrated that
its sanitive effects during hot weather would
be scarcely less than during the cold season.
This statement will excite the surprise and
incredulity of many who have not considered
the subject in all its bearings, nor familiar-

ized themselves with some of the most important inventions and discoveries of the age.

It is well known that hot weather is often quite as injurious to those suffering from pulmonary disease as the cold, changeable weather of winter. Dr. Ramadge of London, in his work on Consumption, stated that the cases of this disease that came under his notice in summer were nearly double the number that he treated in winter. He gives as the cause of this increase the augmented temperature of the weather, increasing the intensity of two of the most important stages of the hectic paroxysm, — the hot and the sweating. Dr. Rush also found the summers of Philadelphia very unfavorable to those having this disease. The mortuary statistics of Massachusetts show that about as many die of lung-disease in summer as in winter.

It is obvious, therefore, from the above

6

and much other evidence that might be
adduced, that those afflicted with pulmonary
disease need a mild, equable temperature all
the year round.   A .temperature never rising
much above 65° Fahrenheit, nor falling
much below that point, would undoubtedly
be the best.   It has already been shown that
such a temperature · could be easily main-
tained in the glass garden during the cold
season ; and it is proposed to show that the
temperature of the air within could be kept
down to that point during the hottest days
of summer.

A few years since a machine was intro-
duced into some of the English collieries to
perform the difficult and dangerous work of
kirving, or cutting under the seams of coal ; *
an operation which had previously been per-

---

* Compressed air was first practically used in the construction and ventilation of the Mont-Cenis Tunnel, and it
has since been used in the Hoosac and other tunnels.

formed by hand, and by which many lives had been destroyed. The machine did the work much quicker, cheaper, and better than it had previously been done; and at the same time it produced another most important benefit, — it cooled and ventilated the mines. It was operated by means of air, highly compressed by an engine at the mouth of the mine, and conducted by flexible tubes to the required spot. According to a well-known law in physics, air, when compressed to a sufficient degree, is deprived of its heat. Tyndall has lately succeeded in compressing it to such a degree, that, when it issued from the pipe, it was so intensely cold as to congeal all the moisture of the room into minute snowflakes. In the collieries it was found that the air used to drive the machines at a pressure of three atmospheres, or somewhat over forty pounds to the square inch, issued from the conducting pipes at nearly a freezing temperature. The oppressive warmth natural to deep mines was very sensibly diminished, and the condition of the air within was also materially improved, by the constant influx of pure cold air from the surface of the ground. Air, artificially reduced in volume, has since been applied to other purposes, such as ventilating long rock tunnels during excavation, and driving the perforating machines.*

---

* "For the construction of the tunnel (Mont Cenis), the great instrument in the hands of the engineers is compressed air. And what an instrument it is! By its aid they furnish air for respiration, wind to drive away vapors, power to run machines; they eject water to play against the rock, produce cold to temper the atmosphere, and heat by a blast at the forges near the entrance. Thus air, wind, power, water, cold, heat, can all be applied, and precisely where they are wanted. This sounds like fable ; but it is a literal truth." — *Report on European Tunnels by* CHARLES S. STORROW, 1862 ; *Massachusetts Reports.*

It is evident, therefore, that compressed air is the easily-applied remedy for too high a temperature in buildings as well as in mines. The boilers in the basements of the iron towers surrounding the sanitarium, which in winter would furnish steam to warm the edifice, would, during hot summer days, furnish the force for working many powerful compressing engines. Pipes of suitable size would lead underground from the towers to all parts of the interior, where the compressed, refrigerated air would escape through large, concealed registers, perforated with small holes like a cullender or the fine rose of a watering-pot. The powerful jets of cold air would thus be so sifted, or divided, that they would escape ordinary notice. Cold air being heavier than warm, it would remain in the lower portion of the garden, where it would do the most good, until warmed by the sun, when it would rise, and flow out through the upper ventilators to make room for a fresh supply. Portions of the roof would be shaded by canvas awnings during hot weather; and the refrigerating effect of the compressed air would be further aided by closing the ventilators in the lower portions of the edifice, while those in the dome and roof would remain open.

By the means above mentioned, the summer temperature of the sanitarium could be kept down to any desirable point. The air within would be perfectly pure and pleasantly cool, like mountain-air in warm weather. The susceptible visitors would encounter none of those sudden atmospheric changes common to our climate at all seasons, and which are quite as injurious in their effects upon invalids in summer as in winter. So far as the important conditions of temperature and pure air could aid in the recovery of the visitors, they would be far better situ-

ated than they could be in any natural
climate to be found on the globe. Their
surroundings also, social, entertaining, sanitive, and educational, would far surpass any
other summer-resort in the country. Invalids,
however querulous and notional some of
them may often appear, are still an intelligent class: they would not be backward in
recognizing, and availing themselves of, the
inestimable advantages of such a refuge from
the debilitating heat of the dog-days. The
sanitarium in summer would be quite as well
patronized as during the wintry season.
Like St. Peter's at Rome, it would have its
own climate, independent of the changing
seasons outside. To its equable temperature
would apply the description of Madame de
Staël of the unvarying climate of the
interior of the great Roman basilica: "Il
a ses saisons a lui, son printemps perpetuel,
que l'atmosphere du dehors n'altere jamais."

But the perpetual spring of St. Peter's, with
its " dim religious light," its cold stone
floor, and its ponderous columns, could bear
no comparison with the cheerful brightness,
the expanded area, and the leafy beauty, of
our ideal garden-palace.

Such is a general description of the author's
plan of a sanitarium, on a scale commensurate in some degree with the importance of
the end to be subserved, but not all beyond
the resources of any large civilized community to carry out.   Even if all the details of
the plan were perfected (which, it is needless
to say, they are not), they could not well be
given in an essay intended to be untechnical
and popular in its scope.   Many desirable
minor features of the plan, that have recommended themselves to the approval of the
writer during his study of the subject, have
not been mentioned.   For instance, the walls
inside of the central garden, as well as of the

arcade and connecting passage-ways, could
be utilized to advantage by training up the
supporting columns and mullions thousands
of vines of the Hamburg, Chasselas, Muscat,
and other fine varieties of foreign grapes
which will thrive in this climate only under
glass. Great quantities of the finest fruit
could be grown and ripened in this way
without extra expense, which, with a little
care in keeping, would supply the hotel-
tables throughout the winter with grapes for
the dessert, contributing in no small degree
to the health and gratification of the visitors.
Another plan of utility, which would recom-
mend itself especially to those many visitors
who were lovers of fancy poultry, would be
to use one or more of the open-air gardens,
between the central edifice and the arcade,
for extensive poultry-yards. In these large,
sheltered, sunny ranges, numerous flocks of
the best breeds of gallinaceous and aquatic

fowls would help to furnish eggs and chick-
ens for the *tables d'hôte;* and, besides, would
be a pleasing addition to the amusements of
the guests.   But many such details as these
must be left till the capital is subscribed, a
board of directors chosen, a location pur-
chased, and the ground-plan and elevation
of the buildings decided upon.

CHAPTER III.

CONCLUSION.

THERE is a homely Scotch proverb, the sentiment of which Emerson has finely elaborated in his essay on "Compensation," which says, "There was never a stinging nettle that hadn't a dockin-leaf close beside it:" the moral deducible from which is, that, for nearly all the evils of this world, a remedy is to be found close at hand.

The dockin-leaf that is to cure the wounds caused by the stinging nettles of harsh climates is the glass pane, aided by other appliances, some of which have been mentioned in the preceding pages. It needs no belief in a future millennium to be very cer-

tain that the years of the world yet to come
will see a greater regard for human life, and
a more earnest effort to prevent its needless
waste, than prevails at present.   As it is, man
does not live out half his allotted days.
If we accept the deductions of M. Flourens,
the eminent French *savant*, then a person
who has reached the threescore and ten
years of the Psalmist ought to be scarcely
beyond the prime of life.   Flourens' investigations into the laws governing the duration
of animal life, though perhaps not entirely
reliable, have yet a certain interest; and his
conclusions in regard to the natural limit of
human life may be correct, even should the
premises on which they were based prove to
be untenable.   He found, that, as a rule, the
inferior animals live on the average about
five times as long as the time it takes them
to attain their full growth.   Thus a horse is
about five years in getting his growth; and

the average age of horses is not far from twenty-five years: and so of other animals, domestic and wild. Reasoning from analogy, Flourens concluded, that, as the same laws of being govern mankind that rule the destinies of horses and other animals, and as it requires twenty years for man to get his growth, then the average duration of human life should be five times twenty, or one hundred years. "A man between sixty and seventy," says the ingenious Frenchman, "ought to be only in the full maturity of his powers, physical and intellectual."*

Whether this hypothesis of the natural limit of human existence is the true one or not, it is evident that some do attain a life of one hundred years and upwards; and perhaps all would, if inherited tendencies and sur-

---

* For a full exposition of this theory of M. Flourens, see his work entitled "De la Longévité Humaine, et de la Quantité de Vie sur le Globe."

rounding circumstances were equally favorable to length of days. But sanitary science will have to make long strides before centenarians become very plenty in our streets. Before that far-off time arrives, consumption, which sweeps off the very flower of human-. ity at the present day, will be so far conquered as to demand a much less proportion of the victims of the grim destroyer than one out of five. As some of the readers of these pages may be inclined to doubt the curability of this disease, even in its earlier aspects, it will not be out of place to quote the opinion of the celebrated Laennec, one of the most skilful physicians that ever lived, and who made the study of pulmonary disease a specialty.

" The cure of consumption,"* says Laen-

* "La guérison daus les cas de phthisie pulmonaire où l'organe n'a pas été entièrement envahi, ne presente, ce me semble, aucun caractère d'impossibilité, ni sous le rapport de la nature du mal, ni sous celui de l'organe affecté."

nec, " when the lungs have not been entirely disorganized, ought not to be considered at all impossible,' either as regards the nature of the disease or the part affected. The destruction of a part of the substance of the lungs is not necessarily mortal, since even wounds of this organ are frequently healed."

Dr. Caswell, the almost equally eminent English physician, adds his testimony to that of his French *confrère*. He emphatically says, —

" We cannot avoid repeating the fact, that pathological anatomy has, perhaps, never afforded more conclusive evidence in proof of the curability of a disease than it has in that of tubercular phthisis." Testimony of the same tenor from other distinguished physicians might be cited, if this were meant to be an exhaustive treatise on pulmonary disease. But it will suffice, for the purposes of an untechnical essay, to bring forward two

such names as the above in support of the
position that consumptive disease is not
necessarily fatal ; that, unless too deeply
rooted, it will yield to the healing forces of
nature, if the proper conditions of air, temperature, exercise, and diet, are furnished to
the patient. It has been the object to show,
in the previous chapters, that these conditions could be found in perfection nowhere
else than in a sanitarium like or resembling
the one described. Such an establishment
would be under the charge of a superintend-
ing physician of the highest intelligence and
the strictest integrity, assisted by a corps of
subordinates selected for the same qualities.
All the means that have proved beneficial
in the treatment of the malady, such as concentrated nutriment, vocal gymnastics, in-
spiration and expiration of the breath
through tubes, &c., would be discriminatingly
used to assist Nature in her efforts to throw

off the disease. As to prevent is always better than to cure, those persons who were attacked with the symptoms of incipient phthisis would find the sanitarium a ready refuge, where all such unfavorable symptoms would speedily vanish under the genial influences of the place. If such an establishment were under the control of the State, and if it were made compulsory upon physicians to report every case of incipient lung-disease to the proper authorities as is now the case in regard to small-pox, then the immediate removal of such persons to the sanitarium would insure their speedy recovery; and thus the dragon of Consumption, having no victims to feed upon, would, with a sort of poetic justice, die of atrophy. The philanthropist, the capitalist, or the legislator, who shall aid in such a desirable result, will deserve, far more than that old knight of Malta for his battle with the python, to

7

have the title " Draconis Extinctor " engraved
upon his monument.

In the preceding pages some hard things
have been said about our American climate,
which were, nevertheless, quite within the
bounds of truth. But equally hard things
have been said of other climates. None of
the earth climates is perfect in all respects.
The writer once saw a weather-diary that
had been kept by a person of a somewhat
sensitive, poetic nature. In it he had noted
down all the days in the year when the
weather seemed to him to be perfect, — days
when it was neither too hot nor too cold,
neither too wet nor too dry, too windy nor
too cloudy. The number of such perfect
days, according to his reckoning, amounted
to just five out of the three hundred and
sixty-five! Perhaps other years would show
a better record, and a less fastidious observer
might have recorded more days that were

perfect in the same year. To a person full-blooded and in vigorous health, the weather, save as it affects his material interests, is a matter of little concern. To such a one, exposure to the bracing, wintry north-westers exhilarates and tones the whole system; and a poetic temperament, even in a frail physical organization, will find a keen delight in threading woodland paths while the wintry gales are sweeping through the tree-tops overhead. What Goethe said of youth, that it is intoxicated without wine, might be often said at such times of middle age, if not of senility. But, after making all due allowance for the good points which our American climate undeniably has, the fact still remains, that a great majority of the days of the year are not as agreeable nor as healthful to people of sensitive organizations — which term includes most women as well as many men — as they might be. It will be one of the

chief objects of the social and sanitary
science of the future, to neutralize, as far as
possible, these defects of climate.

In submitting to the public the foregoing
plan of an establishment adapted either for
a sanitarium or for a pleasure-resort, it is
not claimed that no improvements can be
suggested in the more unimportant features;
but the central and main ideas here advanced are believed, for the best of reasons,
to be impregnable to the assaults of hostile
criticism. Probably there are some good
matter-of-fact people, who are perfectly
satisfied with the world as it is, who will
consider the whole thing an idle dream of
the imagination, about as unsubstantial in
basis as the "stately pleasure-dome" in
Kubla Khan; but such incredulous souls are
respectfully reminded that all things have to
be imagined before they can become accomplished facts in the world of realities. To

create any thing of importance before it was imagined, or ideally created in the mind, would be an obvious impossibility. Those who assume to decide that any proposed undertaking is impracticable at the present day are very liable to have it conclusively shown that their souls are not prophetic ones. Such was the case with Dr. Dionysius Lardner, whose pamphlet, written to prove that a steamship could never cross the Atlantic Ocean, first reached this country by way of a steamship.

It is to be expected that some of the ideas advanced in these pages will be assailed and criticised. That is the ordeal which all ideas, with any claim to novelty or importance, have to encounter. But if they have, as is believed, a firm foundation in truth and reason, they will not be put down by cavil or vapid sarcasm. Some of the objections may seem plausible at first to those who have not fully considered the subject, while others will have as little relevancy as the objection of a member of the British Parliament to the first railroad that was chartered in Great Britain. "Suppose," said this sapient member of the committee before which the great North-Country engineer was advocating his proposed undertaking, — "suppose that a cow should get on the track of your railroad when the engine was coming along." — "So much the worse for the coo," was the terse reply.

Of not much more cogency than the committee-man's objection to the railroad is the one that the glass in the garden-palace would be liable to be broken by summer hail-storms.* If hailstones of ordinary size

---

* Another objection of even less cogency is the one, that a heavy fall of snow would crush in the roof of the edifice. Perhaps an expert mathematician — Prof. Pierce, for instance — might be able to calculate for what fraction

should fall on glass one-third of an inch or more thick, it would be so much the worse for the hailstones. As for those extremely rare cases when balls of ice of great size fall from the clouds, they are so infrequent, that they may be considered out of the range of ordinary probability. Such a meteorological bombardment, depending, as the iceballs do for their formation, upon certain uncommon electrical states of the upper air, is one of the rarest of natural phenomena, and is always very limited in its range. The chances that an ice-storm, severe enough to break glass of the thickness proposed, would pass over any given locality once in a century, could not be greater than one in a million. During hot weather, large portions

---

-of a second a snow-flake would remain as a snow-flake on the garden-roof. Snow would melt almost as soon as it touched the glass, warmed from below; and would pass off by the proper conductors as rain-water.

of the garden-roof would be covered with canvas, which, as far as it extended, would be a'protection against hail; and, if it were thought advisable, the whole could be completely shielded by covering it with a netting of small galvanized wire, with meshes an inch or so in diameter.

. Only one severe hail-storm has passed over the Sydenham Crystal Palace since its erection, and then no damage was done. By a most singular coincidence, it rattled down upon the glass roof with tremendous din while the grand Hailstone Chorus from Israel in Egypt was being performed, at a musical festival in honor of Handel, by an immense chorus of singers and a large orchestra.

The question now arises, Who among the wealthy, the philanthropic, the men of great. business-energy and of far-seeing minds, will aid in furnishing the required capital for an initial establishment like or resembling the one inadequately described in the preceding chapter? There are a few men in the country who could spare the amount needed for its realization, and scarcely miss it from their colossal fortunes. But the money would not be sunk nor thrown away. If there is any reliance whatever to be placed on calculations *à priori*, such an outlay of capital would be remunerative to its owners, as well as a great public benefit. What an opportunity to secure undying fame for some money-king who has the far-sightedness and the audacity to embark a portion of his accumulated millions in such an enterprise! Is there not one such man in the country, endowed with the sagacity, the liberal-mindedness, and the nerve, to boldly take the initiative, and show the world what may be done by wealth when directed by a high purpose? Or must it be left for the co-operation of men of smaller means? State or national aid is not to be expected, until public opinion, which gives law to the law-makers, is educated to see the importance and the feasibility of the enterprise. If it were possible — and why is it not? — to enlist the interest of a few men endowed with the persevering energy displayed by some of the master-spirits who have carried to successful completion the great undertakings of the age, one such establishment might be made ready for dedication on the approaching centennial of the nation's birth.

It may be that the present generation of men will not see the creation of even one such oasis of bloom and warmth amid the cheerless ice-deserts of our Northern winters. If, unfortunately, that should be the case, then we shall have missed a practical realization of the beautiful possibilities of existence in artificial climates on this weather-beaten section of the earth.

APPENDIX.

The almost unlimited potentiality of glass in the amelioration of the winter climates of high latitudes will yet be generally recognized. There are many ways by which its more extensive use would render life during the cold season greatly more endurable. How easily, for example, and at what comparatively small expense, could the bleak, storm-swept streets of Northern cities in winter be converted into delightful thoroughfares, as far as pedestrianism is concerned, by enclosing the sidewalks with large, thick glass panes, supported in a light, ornamental iron framework! No alteration whatever in the buildings or the streets would be needed. Small iron columns, four or five feet apart, would rise from the curb-stones to nearly the height of the first story of the stores or dwellings. These columns would support a light, open-work entablature, from which the glass roof would slope upwards to the side of the buildings. The spaces between the columns could be left square at the top, or made more ornamental by arches, either round or Gothic. At the bottom, for three feet or upwards in height from the curbstones, iron plates would be substituted for glass, to avoid accidental breakage from hubs of wheels or other causes. Openings would be left at the ends of blocks, where there were cross-streets, suffi-cient to allow free passage to the tide of foot-people. There would also be openings opposite each store or house front. The glass in the sides would be set in movable iron sashes, which would be taken out on the approach of warm weather, and replaced before winter. The glass in the roof would remain permanently throughout the year, and in summer would be covered with canvas awnings.

The cost of thus enclosing the sidewalks of the principal streets in any large city would not be great; while the benefits that would follow, in the promotion of health and comfort, would be incalculable. The thousands of pedestrians that daily throng any great city thoroughfare would, if its sidewalks were thus enclosed, be almost entirely sheltered during the inclement season from storms, cold winds, and dust : they would also, to a great degree, be exempt from the annoyances of snow and mud and the dangers of icy flag-stones. Ladies and delicate invalids could do their shopping or visiting, or take their needful daily exercise, in all weathers. Nor would the benefits of these arcades end with the cold season. The awning-covered roofs would in summer be a protection from rain, from the hot noonday sun, and, to some extent, from dust. Only one small interest will suffer when this entirely feasible, and certainly desirable, plan shall be generally adopted ; and that is the umbrella-makers. But there will still be a demand for their useful commodities from the country-towns.

It admits of no question that the business-street

or block, in any considerable town, which shall
be the first to have glass arcades for the protection of its. sidewalks during the wintry season,
will attract trade enough to its stores to pay the
cost of such enclosures many times over. It is
one of the anomalies of our civilization, that,
while the greatest care is justly taken to prevent
needless suffering among domestic animals, no
protection whatever, save a few slight awnings
in summer, is provided for the throngs of men,
women, and children, that daily pass and repass
along the sidewalks of every large town. They
are exposed to the full peltings of the fierce wintry storms ; to frequent danger to limb, and even
life, by the often icy flaggings ; and, in the hot
season, to the scorching noonday sun and the
frequent torrents of rain.

In view of this neglect to provide the means
conducive to human health and comfort in city
streets, would it not be well to start a society for
the prevention of cruelty to pedestrians, with
some philanthropic Mr. Bergh for its president,

the aim of which should be to agitate the matter in all reasonable ways until the proper remedies were provided? The city of the future will undoubtedly have better appliances for the comfort and well-being of its inhabitants than do the cities of to-day.

As only by often-repeated iteration do unfamiliar ideas become familiar and accepted ones, it will not be out of place to restate, in conclusion, the main points advanced in the foregoing exposition. They are briefly summed up as follows: that while vicissitudes of climate cannot be controlled by human agency, yet it is possible to neutralize, to a great degree, their ill effects; that the three principal agents, which in the near future will be used to make this possibility a reality, are GLASS, STEAM, and COMPRESSED AIR. By the intelligent use of these means, the cold of winter and the heat of summer may each be neutralized, and artificial climates be created, agreeable and healthful in temperature, and, if desired, nearly invariable throughout the year.

www.ingramcontent.com/pod-product-compliance
Lightning Source LLC
Chambersburg PA
CBHW022340020726
47500CB00004B/1215